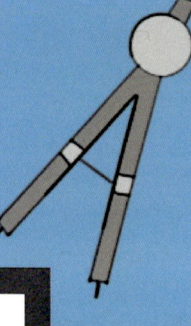

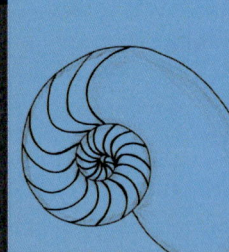

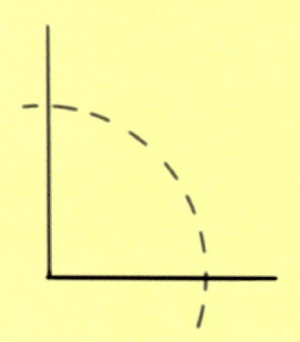

# MATH MARVELS:
## 30 Cool facts to WOW your Brain!

Written by
**Taarun Krushanth**

# HI...

I'm Math kid!

I'm 6 years old! At the time this book was published, anyway.

I know what you're thinking... I'm too young to write a book about Math.

It might surprise you to know that right now I'm learning Statistics and Probability and from there... who knows!

I **LOVE** math! It's more or less all I talk about.

Sooooooo.... I thought why not write down some interesting math for other kids to inspire their math journey.

I might even inspire some grown-ups like....

...MY MUM!

# SHE IS AMAZING!

She helps me by teaching and sometimes learning with me!

She finds resources and takes me to the library to get interesting books and finds fun math related activites for me!

She listens to me all day and sometimes all night! (She says I even do math in my sleep!)

And...

# MY AUNTIE

She isn't math-sy but she listens to me and even lets me teach her.

She helped me design this book!

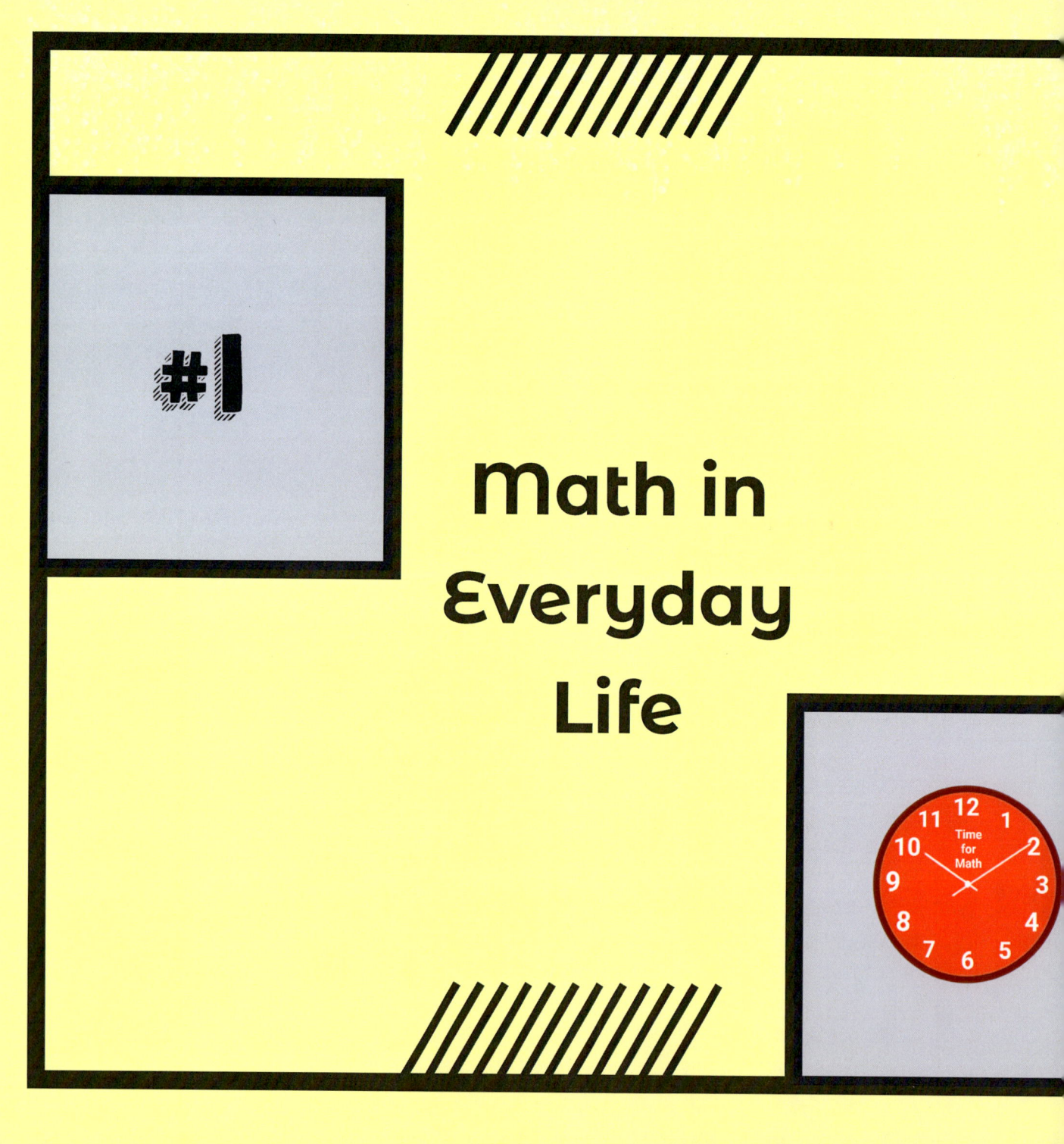

# Let's dive deep into some examples of how we use math in everyday life:

**Playing Games:** When you play board games like Monopoly or card games like Uno, you count the spaces you move or the points you score. In sports, you keep track of scores to see who's winning. For example, in soccer, you count how many goals each team has.

**Shopping:** When you go shopping, you use math to add up the prices of the items you want to buy. You also use math to calculate discounts. For example, if a toy is 20% off, you can figure out how much money you save.

At the cash register, you check if you have enough money to pay and calculate the change you'll get back.

**Cooking and Baking:** Recipes use measurements like cups, teaspoons, and grams. If a recipe says you need 2 cups of flour, you measure it out to get the right amount. Sometimes, you need to double a recipe or cut it in half. Math helps you figure out how much of each ingredient you need.

**Traveling:** When going on a trip, you might use math to calculate how long it will take to get there. If you're riding a bike or walking, you might measure the distance and figure out the best route.

**Telling Time:** You use math to read the clock and figure out how many minutes until your favourite TV show starts or how long you have to finish your homework.

Understanding time also helps you manage your day, like knowing when to leave for school or how long you can play before bedtime.

*Math helps us understand the world around us and make decisions. It's like a superpower we use everyday!*

"Let's bake something!"

# Math & Science

Let's explore how math and science work together like best friends to help us understand the world.

## Math Helps Science

Math is like a toolbox for scientists. When scientists want to understand how things work, they use math to measure, calculate, and make sense of what they find.

**Measuring and Counting:** If a scientist wants to know how tall a tree is, they measure it using numbers. If they want to know how many birds live in a forest, they count them.

**Experiments:** Scientists do experiments to test ideas. They use math to collect data (like how fast something moves or how much something weighs) and figure out what it means.

**Making Predictions:** Scientists use math to make predictions. If they know how fast a rocket travels, they can calculate when it will reach space.

## Science Helps Math

Science helps math by giving it real-world problems to solve. Without science, math would just be a bunch of numbers and rules without a purpose.

**Physics and Math:** Physics is a science that studies how things move and work. It uses a lot of math. For example, math helps physicists understand gravity, which is why we stay on the ground and don't float away.

**Biology and Math:** Biology is the study of living things. Math helps biologists understand things like how fast plants grow, how many animals live in a certain area, or how diseases spread.

**Medicine:** Doctors and scientists use math to figure out the right amount of medicine to give to patients. They also use math to study how diseases spread and find ways to stop them.

Meteorologists, who study weather, use math to predict if it will rain, snow, or be sunny. They measure temperature, wind speed, and more to make accurate forecasts.

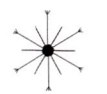

## Fun with Math and Science

Think about baking cookies again. Science explains how ingredients like baking powder make the dough rise, and math helps you measure the right amounts so your cookies turn out delicious. By working together, math and science make baking fun and tasty!

So, math and science are like best friends that help each other understand the world. Math gives science the tools to measure and predict, and science gives math real-world problems to solve. Together, they make amazing discoveries and help us in our everyday lives!

SCIENTISTS USE MATH TO CALCULATE THE PATH OF ROCKETS AND SATELLITES. WITHOUT MATH, WE COULDN'T SEND ASTRONAUTS TO THE MOON OR EXPLORE MARS.

# PYTHAGORAS

#3

The Pythagoras theorem states that the square of the length of the hypotenuse is equal to the sum of squares of the lengths of other two sides of the right-angled triangle.

## LET'S THINK

### Two Dimensions (Flat Surface)

Imagine you have a big piece of paper. You draw a right-angle triangle on it, which is a triangle with one corner that is perfect square corner (like the corner of a book). If you know the lengths of the two shorter sides of the triangle, you c find out how long the longest side (called the hypotenuse)

### THE RULE IS:

$\{\text{Longest side}\}^2 = \{\text{One side}\}^2 + \{\text{Other side}\}$

So, if one side is 3 units long and the other is 4 units long, you d

$$3^2 + 4^2 = 9^2 + 16 = 25$$

Then, the longest side is:

$$\sqrt{25} = 5$$

# ABOUT IT LIKE THIS:

## Three Dimensions (Space)

Now, imagine you are building something with blocks. You want to find out the distance from one corner of a box to the opposite corner, going through the inside of the box. This box has three sides: length, width, and height. The rule for the distance across the box is:

$$Distance^2 = Length^2 + Width^2 + Height^2$$

So, if the box is 3 units long, 4 units wide, and 5 units tall, you do:

$$3^2 + 4^2 + 5^2 = 9 + 16 + 25 = 50$$

Then, the distance across the box is:

$$\sqrt{50} = \text{approx. } 7.1$$

## Higher Dimensions (More Complicated Spaces)

Imagine you have even more directions to move in, like a super complicated video game with many levels. The same kind of rule works to find distances. You just add up the squares of all the sides you are interested in.

So, in any number of dimensions, if you know the lengths of the sides, you can find the distance by using the same idea: add up the squares of the lengths of the sides, and then take the square root to find the distance.

IT'S LIKE MAGIC, BUT IT'S REALLY JUST MATH THAT WORKS IN EVERY KIND OF SPACE!

# Exponential Growth

Let's explore exponential growth using a fun story about grains of rice and a chessboard.

## The Chessboard and the Grains of Rice

Imagine you have a chessboard. A chessboard has 64 squares (8 rows and 8 columns). Now, let's play a game with grains of rice:

Start with One Grain of Rice:

Place one grain of rice on the first square.

Double the Rice on Each Square:

On the second square, place two grains of rice.

On the third square, place four grains of rice.

On the fourth square, place eight grains of rice.

Each time you move to the next square, you double the number of grains of rice from the previous square.

Let's see how the number of grains grows:

1st square: 1 grain

2nd square: 2 grains (1 x 2)

3rd square: 4 grains (2 x 2)

4th square: 8 grains (4 x 2)

5th square: 16 grains (8 x 2)

As you keep going, the number of grains on each square gets very big, very fast.

Exponential Growth

This kind of growth, where you keep doubling, is called exponential growth. It means things grow really, really fast.

## Why is it Amazing?

Let's see how many grains are on some of the later squares:

10th square: 512 grains

20th square: 524,288 grains

30th square: 536,870,912 grains

By the time you get to the 64th square, the number of grains of rice is so huge, it's hard to imagine!

## Why It's Important

Exponential growth shows how quickly things can add up when they keep doubling. It's like magi

Real-Life Examples

**Populations:**

If bacteria multiply by splitting into two every hour, their numbers grow very fast, just like the grains of rice.

**Computers:**

Computer speeds and storage capacities have grown exponentially over the years, meaning they've gotten much faster and can hold a lot more information very quickly.

# Fun Fact

If you actually placed all those grains of rice on the chessboard, you'd end up with more rice than there is in the entire world! That's how powerful exponential growth is.

So next time you see something growing very quickly, remember the story of the rice and the chessboard. It helps us understand why things that grow exponentially can get very big, very fast!

# #5 Googology

10,000,000,000,000,000,000,000,000,000,000,000,000,000,000,000,000,000,000,000,000,000,000,000,000,000,000,000,000,000,000,000,000,000,000

**Googology is the study of large numbers**

Have you ever wondered how big numbers can get? Like, REALLY big numbers? Well, that's what Googology is all about! Googology is the study of really big numbers. These numbers are so huge that they make your head spin just thinking about them!

## Where Did It Start?

It all started with a number called a "googol." A googol is a 1 followed by 100 zeros. That's a lot of zeros! Here's what a googol looks like:

**10,000,000,000,000,000,000,000,000,000,000,000,000,000,000,000,000,000,000,000,000,000,000,000,000,000,000,000,000,000,000,000,000,000,000**

Can you imagine counting to that number? It would take forever!

The name "googol" was actually made up by a 9-year-old boy named Milton Sirotta in 1938. Pretty cool, right?

## Even Bigger Numbers

If you think a googol is big, wait until you hear about a googolplex! A googolplex is a 1 followed by a googol zeros. That's like trying to write a googol of zeros! It's so big that if you tried to write it out, there wouldn't be enough space in the entire universe to fit all the zeros!

## Why Do We Study Big Numbers?

You might wonder, "Why do we need such big numbers?" Well, studying big numbers helps mathematicians and scientists understand the limits of what we can count and measure. It also helps them think about really large concepts in science, like the size of the universe or the number of possible chess games.

## Graham's Number

One of the biggest numbers ever used in a serious math problem is called Graham's number. It's so gigantic that even if you wrote it out, your brain wouldn't be able to comprehend it. Instead of writing it down, mathematicians use special notation to describe it. It's so big that if you tried to think about it, your head might feel like it's going to explode!

## Playing with Big Numbers

You can have fun with big numbers too! Try making up your own super huge number and give it a funny name. Maybe you could invent the "tenyojillion" or the "gigantosaur." The possibilities are endless!

## Wrapping Up

So, next time you think about numbers, remember that they can get unimaginably big. Googology is like a playground for numbers, where they can grow as large as your imagination can take them. Who knows? Maybe you'll be the next person to come up with an even bigger number!

# #6 INFINITY

∞

## Infinity is a concept that describes something that never ends.

Imagine the biggest number you can think of. Got it? Now, add one to it. You can always add one more, right? No matter how big the number is, there's always a number bigger than it. That's infinity! It's like a never-ending adventure with numbers.

Here's a fun way to think about it: Picture a giant cookie jar. No matter how many cookies you take out, there are always more cookies inside. The jar never runs out of cookies. That's what infinity is like with numbers—they never run out!

Infinity is a special kind of number because it's not really a number you can count to. It's more like an idea that helps us understand things that go on forever. For example, the number of stars in the sky, the grains of sand on a beach, or even the time you could spend counting numbers.

So, infinity is a super cool concept because it reminds us that some things are just too big to measure, and the possibilities are endless!

Did you know there are different types of infinity?

It might sound surprising, but not all infinities are the same size. Let's explore this idea with a simple story.

Imagine you have two friends, Alice and Bob. Alice loves to count whole numbers (1, 2, 3, 4, ...), and Bob loves to count even numbers (2, 4, 6, 8, ...)

Alice's Infinity:

Alice starts counting: 1, 2, 3, 4, and so on. She will keep counting forever because there are infinitely many whole numbers.

Bob's Infinity:

Bob starts counting even numbers: 2, 4, 6, 8, and so on. He also will keep counting forever because there are infinitely many even numbers.

Now, you might think that both Alice and Bob have the same amount of numbers to count, right? They both have infinity. But here's the interesting part: mathematicians have discovered that some infinities are actually larger than others!

Bigger Infinity:

To see a bigger infinity, let's think about all the numbers between 0 and 1. These include 0.1, 0.12, 0.123, 0.1234, and so on. You can keep adding more and more digits after the decimal point, and there are infinitely many possible numbers between 0 and 1.

So, we have:

- Alice's whole numbers: 1, 2, 3, 4, ...

- Bob's even numbers: 2, 4, 6, 8, ...

- And all the tiny numbers between 0 and 1: 0.1, 0.12, 0.123, 0.1234, ...

There are actually more numbers between 0 and 1 than there are whole numbers or even numbers. This means that the infinity of numbers between 0 and 1 is a "bigger" infinity than the infinity of whole numbers or even numbers.

# Prime Numbers

Let's talk about prime numbers in a fun and simple way.

Imagine you have a box of toys. Some toys can be grouped in many different ways, but some can only be grouped in one special way. Prime numbers are like those special toys that can only be grouped in one particular way.

## Here's how it works:

A prime number is a number that can only be divided evenly by 1 and itself.

This means that when you try to split a prime number into equal groups, the only groups you can make are just one big group of the whole number or individual groups of 1.

The number 2 is a prime number. You can only divide it into 2 (2 divided by 2 equals 1) and 1 (2 divided by 1 equals 2). No other numbers work.

The number 3 is also a prime number. You can only divide it into 3 (3 divided by 3 equals 1) and 1 (3 divided by 1 equals 3).

Other examples of prime numbers are 5, 7, 11, and 13. These numbers can only be divided evenly by 1 and themselves.

## Why are Prime Numbers Special?

Prime numbers are like the building blocks of all numbers. Just like you can use blocks to build many things, you can use prime numbers to create other numbers.

- For example, you can multiply the prime numbers 2 and 3 to get 6. Or you can multiply 2, 2, and 3 to get 12.

## Not Prime Numbers

Some numbers are not prime because you can split them into different equal groups. For example, 4 is not a prime number because you can divide it by 1, 2, and 4. (4 divided by 2 equals 2).

The number 6 is also not a prime number because you can divide it by 1, 2, 3, and 6.

**In summary, prime numbers are special numbers that can only be split into one big group of the whole number or individual groups of 1. They are like the unique toys in your box that can only be grouped in one special way.**

# Fermat's Last Theorem #8

Imagine you have a big puzzle with numbers, like a jigsaw puzzle but with numbers instead of pictures. Fermat's Last Theorem is about a very specific type of number puzzle.

## Here's what the theorem says:

When you learn math, you often see equations like $a^2 + b^2 = c^2$. This means you add two squares of numbers to get another square. For example, $3^2 + 4^2 = 5^2$ because $9+16 = 25$.

Fermat's Last Theorem says that if you change the equation to use powers higher than 2, like $a^3 + b^3 = c^3$ or $a^4 + b^4 = c^4$, then there are no whole numbers (positive numbers like 1, 2, 3, etc.) that can make the equation true.

In other words, if you try to find three whole numbers a, b, and c that fit this new equation, you won't be able to.

This might seem simple, but proving it was really, really hard. Mathematicians tried to solve this puzzle for over 350 years!

A mathematician named Andrew Wiles finally solved it in 1994. He worked on this puzzle for many years and used a lot of very advanced math to prove that Fermat was right.

So, Fermat's Last Theorem is like a very tricky puzzle with numbers that says, "If you try to add two cubes (or higher powers) of numbers to get another cube (or higher power), you can't do it with whole numbers." And it took a very long time for someone to solve this puzzle!

//#9

# The Adventure of Geometry

Geometry is the study of shapes, sizes, and spaces. It's like going on an adventure to explore the world of forms and figures.

Geometry is the study of shapes, sizes, and spaces. It's like going on an adventure to explore the world of forms and figures.

Angles can be:

Right angle: 90 degrees, like the corner of a square.

Acute angle: Less than 90 degrees, like a sharp corner.

Obtuse angle: More than 90 degrees, like a wide corner.

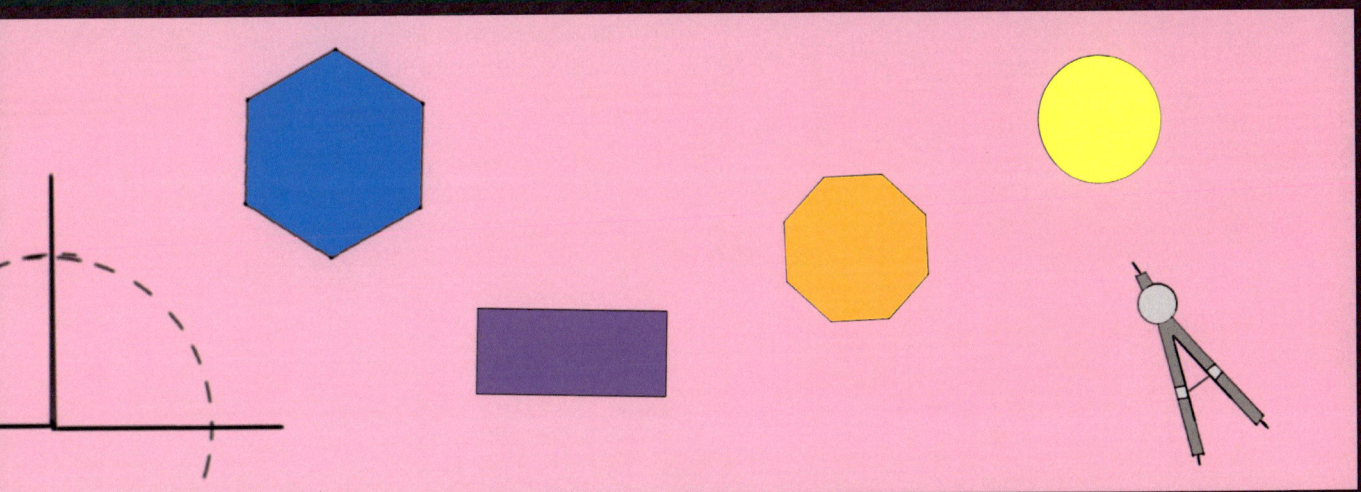

## Perimeter and Area

Geometry helps us measure things. The perimeter is the distance around a shape. For a rectangle, add up all the sides. The area is the space inside a shape. For a rectangle, multiply the length by the width.

## Geometry in the World

Geometry is everywhere! Architects use it to design buildings, artists use it to create beautiful pictures, and even nature uses it to form honeycombs and spider webs.

Geometry helps us see the world in a new way and understand how everything fits together.

# Fractals

## What Are Fractals?

Fractals are special patterns that look the same no matter how close you zoom in.

Imagine you have a picture of a tree. If you zoom in on a branch, it looks a bit like the whole tree.

If you zoom in even more on a smaller branch, it still looks like the whole tree. This repeating pattern is what makes something a fractal.

### Let's start with a simple example called the Sierpinski Triangle.

1. Big Triangle: Start with a big triangle.

2. Smaller Triangles: Draw a dot in the middle of each side and connect these dots to make four small triangles.

3. Remove the Middle Triangle: Take out the middle triangle.

4. Repeat: Now, do the same thing to the three remaining triangles. Draw dots in the middle of each side, connect them, and remove the middle triangle.

You keep repeating this process, and each time you get a pattern that looks like lots of little triangles making up the big triangle. No matter how much you zoom in, you see the same pattern.

### Fractals in Nature

Fractals aren't just in math; they're everywhere in nature! Here are a few examples:

- Broccoli: Look at a piece of broccoli. Each small piece looks like a tiny version of the whole broccoli. Just because they are yucky doesn't mean they can't be mathematically interesting!

- Snowflakes: Snowflakes have intricate, repeating patterns that look similar at different scales.

- Coastlines: If you look at a map of a coastline, it looks jagged and irregular. Zooming in on a small part of the coastline still shows a similar jagged pattern.

## Why Fractals Are Cool in Math

Fractals are fascinating in math because they help us understand complex shapes and patterns. They show us that simple rules can create incredibly detailed and beautiful designs. Here's why they're important:

**Infinite Detail:** Fractals can have an infinite amount of detail. No matter how much you zoom in, there's always more to see.

**Self-Similarity:** Fractals look the same at different scales. This property is called self-similarity.

**Simple Rules**, Complex Results: Fractals show how simple mathematical rules can create complex and beautiful patterns.

## Fun with Fractals

You can even create your own fractals! Try drawing a simple shape and then repeat a pattern inside it, just like with the Sierpinski Triangle. Keep going as long as you can, and you'll see how a simple rule can make a complex and interesting design.

Fractals are a wonderful way to see how math creates patterns and structures all around us. They help us understand nature, art, and many other things in the world. Plus, they're just plain fun to look at and explore!

Benoit B Mandelbrot is the person who gave fractals their name.

## Joke:

What does the B in Benoit B Mandelbrot stand for?

Benoit B Mandelbrot! It's a fractal!

# #11
# Pi (π) is Infinite

Pi (π) is a very special number that represents the ratio of a circle's circumference to its diameter.

What's amazing about π is that it never ends and never repeats. The first few digits are 3.14159, but it goes on forever!

3.1415926535897932384
5028841971693993751O
5923078164062862O899

## What is Pi?

Pi (π) is a special number used in math to help us understand circles. It is the number you get when you divide the distance around a circle (the circumference) by the distance across the circle (the diameter). No matter how big or small the circle is, this number is always the same.

## The Number Pi

When we write down Pi, it starts like this: 3.14159... and goes on and on. But here's the cool part: it never stops, and it never repeats the same pattern. It's like a story that goes on forever without ever repeating any part of it exactly the same way.

**6264338327**
**5820974944**
**8628034825**

## Why is Pi Infinite?

1. Endless Counting: Think about counting numbers. You can count forever: 1, 2, 3, 4, and so on. There's no end. Pi is a bit like that, but with digits after the decimal point.

2. No Repeating Pattern: Some numbers, like 1/3, have repeating patterns (0.333...). But Pi isn't like that. It doesn't have any repeating pattern no matter how far you go. It's like writing a story with no repeated sentences.

3. Finding Pi: To find the digits of Pi, mathematicians use special formulas and computers. They have found millions of digits of Pi, and it still keeps going without repeating!

## A Fun Way to Think About Pi

Imagine you have a really long piece of string, and you start wrapping it around a circle. No matter how long you keep wrapping, you'll always be able to keep measuring more and more precisely how many times that string goes around the circle. The number you get, Pi, will keep going on and on.

So, Pi is an endless number that helps us understand circles, and its never-ending digits make it one of the coolest numbers in math!

# Zero is an Even Number

What is an Even Number?

An even number is any number that can be divided by 2 without leaving any leftovers. For example, if you have 4 cookies and you want to share them with a friend, you can give each of you 2 cookies, and there are no cookies left over. So, 4 is an even number.

What is Zero?

Zero means you have nothing at all. If you have zero cookies, it means you don't have any cookies to share.

How Do We Know Zero is Even?

Let's see if zero can be divided by 2 without any leftovers.: If you have zero cookies and you want to share them with your friend, you would each get zero cookies. There are no cookies left over, right? So, zero can be divided by 2.

Think of a number line with both positive and negative numbers. Even numbers are usually marked in a different colour. If you look closely, you'll see that zero is right in between the positive and negative even numbers, like -2, -4, 2, 4, and so on.

Since even numbers follow a simple rule (can be divided by 2 without leftovers), zero fits this rule perfectly. Zero divided by 2 is zero, and there are no leftovers.

## Cool Fact!

Every even number has a twin:

2 has -2
4 has -4

Zero has itself because zero is in the middle of all even numbers. It makes zero extra special!

The same is true for odd numbers

# The Number 0.999...

Let's explore how the number 0.999... (with the 9s going on forever) is actually the same as 1 using a fun trick with algebra.

Let's say

$x$ is the number 0.999... (where the 9s go on forever).

So, we can write:

$x = 0.999...$

Multiply by 10:

If we multiply both sides of the equation by 10, we get:

$10x = 9.999...$

Now we have two equations:

$x = 0.999...$ and $10x = 9.999...$

Subtract the first equation from the second:

Let's subtract $x = 0.999...$ from $10x = 9.999...$:

$10x - x = 9.999... - 0.999...$

Simplify the subtraction:

On the left side,

$10x - x$ is $9x$

On the right side,

$9.999... - 0.999...$ is exactly 9 (since the repeating parts cancel each other out).

So, we get: $9x = 9$. Solve for $x$

To find $x$, we divide both sides of the equation by 9: $x = 1$

## Conclusion

We started by saying

$x = 0.999...$ and ended up finding that $x = 1$.

This means that 0.999... (with the 9s going on forever) is actually the same as 1!

It's a neat trick that shows how sometimes numbers can be a little surprising.

#13

# #14

## Fibonacci Sequence in Nature

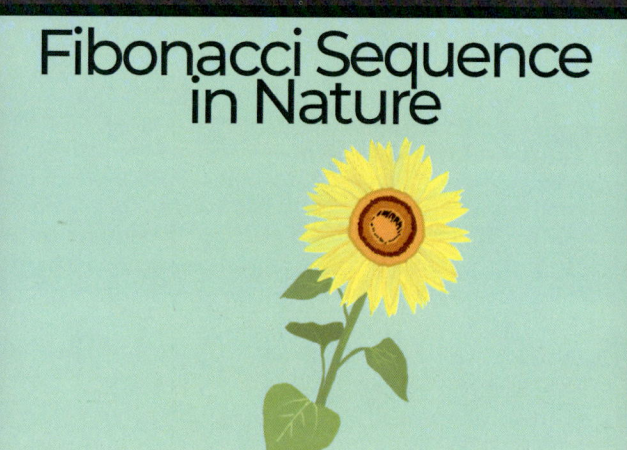

## How Does the Fibonacci Sequence Appear in Nature?

Nature loves the Fibonacci Sequence! Here are some co examples:

### Flower Petals

Many flowers have a number of petals that is a Fibonac number. For example, lilies have 3 petals, buttercups hav 5 petals, and daisies often have 34 or 55 petals.

### Pinecones:

If you look at a pinecone, you'll see that the scales a arranged in spirals. Count the spirals in one direction, ar you'll find a Fibonacci number. Count in the opposi direction, and you'll find the next Fibonacci number!

### Sunflowers

Sunflower seeds are arranged in spirals too. If you cour the spirals, you'll often find Fibonacci numbers. Th pattern helps the seeds pack tightly and efficiently.

### Leaf Arrangements

Some plants grow their leaves in a spiral pattern aroun the stem. If you count the number of leaves in one full tur around the stem, you'll often find a Fibonacci numbe This arrangement helps the leaves get the most sunligh

### Shells

The shells of some snails and sea creatures follow spiral pattern that grows wider in a way described b the Fibonacci Sequence. This helps the animal grov evenly and efficiently.

## What is the Fibonacci Sequence?

The Fibonacci Sequence is a special list of numbers that starts like this:

1, 1, 2, 3, 5, 8, 13, 21, 34, ...

To get the next number in the sequence, you add the two numbers before it:

- 1 + 1 = 2

- 1 + 2 = 3

- 2 + 3 = 5

- 3 + 5 = 8

- And so on...

## Why Does Nature Use the Fibonacci Sequence?

Nature uses the Fibonacci Sequence because it helps plants and animals grow in the most efficient way possible. For example:

- The arrangement of leaves helps plants get the most sunlight.

- The pattern of seeds in a sunflower helps them pack tightly without wasting space

## Fun Activity

You can find Fibonacci numbers in nature yourself!

Next time you're outside, look at flowers, pinecones, or even a snail shell. Count the petals, spirals, or sections, and see if you find a Fibonacci number.

Nature uses these patterns because they help plants and animals grow efficiently and beautifully. It's like nature's secret code for making things work perfectly!

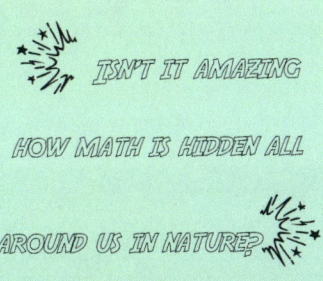

ISN'T IT AMAZING

HOW MATH IS HIDDEN ALL

AROUND US IN NATURE?

# The Golden Ratio

Where Does the Golden Ratio Show Up?

### In Shapes and Art

Imagine a rectangle that is longer and thinner. If you cut off a square from one end, the remaining rectangle will have the same proportions as the original one. That rectangle's length to its width is the Golden Ratio.

### In Nature

- The Golden Ratio shows up in many living things. For example, it can be seen in the spiral pattern of a seashell or the way leaves are arranged on a stem. Even the proportions of your own body, like the length of your arm bones compared to your hand, can follow the Golden Ratio!

### In Architecture

Architects sometimes use the Golden Ratio when designing buildings because it's pleasing to the eye. Ancient Greek temples, like the Parthenon, were built using these proportions.

People like the Golden Ratio because it looks balanced and beautiful. When things are built or designed using these proportions, they often seem just right to us.

Understanding the Golden Ratio helps us see how math and beauty can go hand in hand. You might be surprised where you find it next!

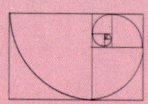

The Golden Ratio is a special number that people have been fascinated with for a long time.

It's about 1.618, and it's often represented by the Greek letter φ (phi).

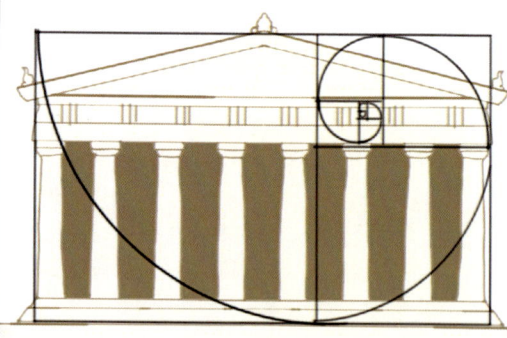

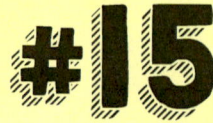

# Hexagons are Efficient

Let's talk about why hexagons are so cool and why bees use them to build their honeycombs.

Imagine you have a bunch of shapes, like squares, triangles, and hexagons.

If you want to cover a flat surface with these shapes without leaving any gaps, some shapes do a better job than others.

Squares and triangles fit together without gaps, but hexagons are even better because they cover the most area with the least amount of edges.

Efficiency:

Bees are super smart when it comes to building their homes, called honeycombs.

They use hexagons because this shape is the most efficient for a few reasons:

Uses the Least Material: Hexagons use the least amount of wax to create their honeycomb cells. This means bees can save their energy and resources.

Strong and Sturdy: The hexagon shape makes the honeycomb strong, so it can hold lots of honey without breaking.

When hexagons are placed next to each other, there are no gaps. This makes the honeycomb very efficient in using space.

Imagine you're drawing a bunch of circles that all touch each other. The spaces in between these circles form a shape, and that shape is a hexagon!

Think of a soccer ball pattern. The patches on a soccer ball are made up of hexagons (and a few pentagons). They fit together nicely without any spaces.

Bees need to store their honey and raise their young in a space that's strong, uses the least amount of resources, and fits together perfectly.

By using hexagons, bees can build a large, strong honeycomb using less wax, leaving more energy for making honey.

So, next time you see a bee or a honeycomb, you'll know that those hexagons are a super smart choice for building a strong and efficient home!

# Logic Gates and Dominoes

#17

Logic gates are like little machines that help computers make decisions. They take in signals (which can be thought of as yes or no, or on or off) and give out a new signal based on some rules.

Imagine you have a bunch of dominoes. If you set them up in a line and push the first one, they'll all fall down, right? We can use this idea to make logic gates!

## AND Gate

An AND gate works like this: it only gives an "on" signal (or lets the dominoes fall) if all the inputs are "on". Imagine two lines of dominoes that meet at a single domino. Both lines need to be pushed at the same time for that meeting domino to fall and continue the chain.

## OR Gate

An OR gate gives an "on" signal if at least one of the inputs is "on". Think of it as having two lines of dominoes leading to a single domino. If you push either line, the single domino will fall.

## NOT Gate

A NOT gate is a bit different. It changes the signal from "on" to "off" or from "off" to "on". Imagine a domino setup where if you push one line of dominoes, it knocks over another line that prevents the final domino from falling. So, if the first line falls, the final domino doesn't; if the first line doesn't fall, the final domino will fall.

By combining these simple setups with dominoes, you can create more complex machines that can solve all sorts of problems, just like how computers work!

Logic gates are closely related to math because they follow specific rules that are part of a branch of math called "Boolean algebra." Boolean algebra is all about working with true and false values, which we can think of as yes and no, or 1 and 0.

## AND Gate and Math

In math terms, an AND gate works like multiplication. If we think of "on" as 1 and "off" as 0:

- 1 AND 1 = 1 (both inputs are "on," so the output is "on")

- 1 AND 0 = 0 (one input is "off," so the output is "off")

- 0 AND 1 = 0 (one input is "off," so the output is "off")

- 0 AND 0 = 0 (both inputs are "off," so the output is "off")

## OR Gate and Math

An OR gate works like addition, but with a twist. In Boolean algebra, 1 + 1 is still 1 because it's like saying "either one or the other or both are true":

- 1 OR 1 = 1 (at least one input is "on," so the output is "on")

- 1 OR 0 = 1 (at least one input is "on," so the output is "on")

- 0 OR 1 = 1 (at least one input is "on," so the output is "on")

- 0 OR 0 = 0 (both inputs are "off," so the output is "off")

## NOT Gate and Math

A NOT gate just flips the value. If it's 1 (on), it turns into 0 (off), and if it's 0 (off), it turns into 1 (on):

- NOT 1 = 0

- NOT 0 = 1

When you set up dominoes to act like these gates, you're creating physical representations of these math rules. By combining these rules, you can make really complex calculations and decisions, just like how a computer does math to solve problems.

SO, BUILDING LOGIC GATES WITH DOMINOES IS LIKE PLAYING WITH MATH IN A FUN AND PHYSICAL WAY!

## #18
## What is e?

The number e is a special number in math, just like π (pi). It's about 2.718, but it goes on forever without repeating.

e is an important number in math, especially in areas like calculus and exponential growth.

**2.71828182845904523535266249775724709366969762772407663035**

## Cool Facts About e

**1. Natural Exponential Growth:**

Think about a tree growing or money in a savings account earning interest. These kinds of things grow in a special way that uses the number e.

**2. Euler's Discovery:**

A Swiss mathematician named Leonhard Euler discovered e in the 1700s. He found it when he was studying how things grow continuously, like how bacteria multiply or how interest on money adds up.

**3. The Magic of Compounding:**

If you put money in a bank, the interest can add up in a magical way. The more often the interest is added (compounded), the closer it gets to growing by the number e

360287471
999595749
3594571382

## Euler's Formula:

There's a super cool equation called Euler's formula that connects e, π, and imaginary numbers. It looks like this:

$e^{i\pi}+1=0$

This equation is amazing because it connects some of the most important numbers in math!

Never-Ending Number: Just like π, e never ends and never repeats. The first few digits are 2.71828, but it keeps going forever.

Fun Example

Imagine you have $1 and you put it in a bank that gives you 100% interest, but they add the interest continuously. After one year, you wouldn't just have $2, you'd have about $2.718 because of the special way e works.

Why e is Special

The number e helps us understand and calculate things that grow or change continuously. Whether it's the way populations grow, how heat spreads, or even how we calculate probabilities in games, e is there making things work smoothly.

So next time you hear about exponential growth or see a cool math trick, remember that the number e is often behind the scenes making it all happen!

Let's play a fun game with numbers and learn something cool about the number e!

## The Game

Get a Random Number Generator:

You can use an online random number generator or write numbers 1 to 100 on slips of paper and draw them from a hat.

Start Generating Numbers:

Start generating random numbers between 1 and 100. Each time you generate a number, add it to your total.

Keep Going Until You Reach 100 or More:

Keep generating and adding numbers until your total is 100 or more.

Count Your Tries:

Count how many numbers it took to reach 100 or more.

## Playing the Game

Let's say you play the game a few times. Here's what it might look like:

First try: You generate numbers 10, 20, 30, 40 (total = 100). It took 4 tries.

Second try: You generate numbers 50, 20, 20, 15 (total = 105). It took 4 tries.

Third try: You generate numbers 5, 10, 15, 30, 40 (total = 100). It took 5 tries.

## How It Works

### Randomness and Averages:

When you generate random numbers, each number can be anything from 1 to 100. The average number you generate will be around 50, but because you stop as soon as the total reaches or exceeds 100, the actual count of tries tends to follow a pattern that relates to e

### More Tries, More Accuracy:

The more times you play the game and average the results, the closer you get to seeing this pattern. It's like flipping a coin many times; the more you do it, the closer you get to an even split of heads and tails.

## Why It's Fun

This game is fun because it combines randomness and a bit of math magic. You get to see a special number, e in action just by playing and counting. Plus, it's a great way to understand how certain patterns and averages appear naturally in numbers.

So, gather your random number generator and start playing!

See how close you can get to e by averaging the number of tries it takes to reach 100 or more. Enjoy discovering the magic of math in a fun and interactive way!

# Number Bases

## What Are Number Bases?

Number bases are like different ways of counting. The most common way we count is called "base 10" (also known as the decimal system), which uses ten digits: 0, 1, 2, 3, 4, 5, 6, 7, 8, and 9. But there are other ways to count, too, using different bases.

## Base 10 (Decimal)

Let's start with base 10 since it's the one we use every day. Here's how it works:

- When you count, you start at 0 and go up to 9.

- After 9, you go back to 0 but add a 1 in front, making 10.

- This pattern continues: 11, 12, 13, and so on.

The place value of each digit in a number tells you how much it's worth. In the number 345:

- The 5 is in the "ones" place (5 ones).

- The 4 is in the "tens" place (4 tens, or 40).

- The 3 is in the "hundreds" place (3 hundreds, or 300).

## Base 16 (Hexadecimal)

Another interesting base is base 16, called hexadecimal. It uses sixteen digits: 0, 1, 2, 3, 4, 5, 6, 7, 8, 9, and then A, B, C, D, E, and F (where A is 10, B is 11, and so on).

- When you count, you start at 0 and go up to F.

- After F, you go back to 0 but add a 1 in front, making 10 (which is 16 in decimal).

- This pattern continues: 11 (which is 17 in decimal), 12, 13, ..., 1F, 20 (which is 32 in decimal), and so on.

## Base 2 (Binary)

Now let's look at base 2, which is called binary. Computers use this system. It only has two digits: 0 and 1.

- When you count, you start at 0 and go up to 1.

- After 1, you go back to 0 but add a 1 in front, making 10 (which is 2 in decimal).

- This pattern continues: 11 (which is 3 in decimal), 100 (which is 4 in decimal), and so on.

The place value works differently in binary. In the number 101:

- The 1 on the right is in the "ones" place (1 one).

- The 0 is in the "twos" place (0 twos, or 0).

- The 1 on the left is in the "fours" place (1 four).

So, 101 in binary is 1 four + 0 twos + 1 one, which equals 5 in decimal.

## How Different Bases Are Used

- Base 10: Used in everyday life for counting and calculations.

- Base 2: Used by computers and digital systems because they operate with two states: on and off.

- Base 16: Used in computing to represent binary numbers more compactly.

## Fun with Bases

Here's a fun way to think about it: Imagine you're playing a game and you can only use a certain number of fingers to count. In base 10, you use all ten fingers. In base 2, you can only use two fingers (0 and 1). In base 16, you get to use your fingers and toes plus some extra digits!

By exploring different bases, you can see how numbers work in various systems, which helps us understand and create amazing things like computers and digital gadgets. It's like having a secret code for counting!

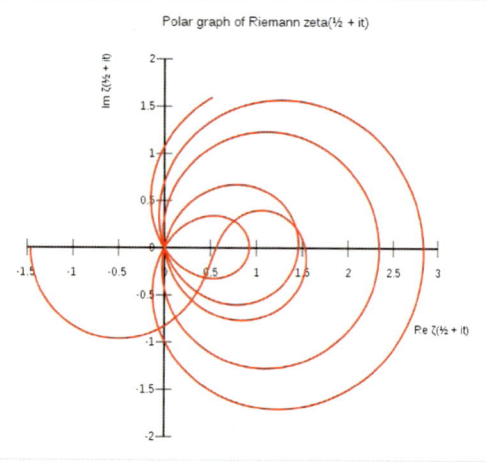

The Riemann Hypothesis is another fascinating puzzle in the world of math. Imagine it like this:

You know about numbers, right? Numbers like 1, 2, 3, 4, and so on. There's a special kind of number called a prime number. A prime number is a number that can only be divided by 1 and itself without any leftovers. For example, 2, 3, 5, 7, and 11 are all prime numbers.

Now, imagine you have a magical number line where all the numbers are lined up. There's a special function called the "Riemann Zeta Function" that helps mathematicians understand the prime numbers and how they are spread out along this number line.

# The Riemann Hypothesis

The Riemann Hypothesis is like a guess about this magical function. It says that there is a special pattern to the prime numbers when you use this function. Specifically, the Riemann Hypothesis suggests that all the interesting points (called "zeros") of this function, where the magic kind of happens, are lined up in a certain way on the number line.

Imagine drawing a straight line through the middle of a piece of paper, and all these special points are supposed to sit exactly on this line. If they do, it means we understand a lot about how prime numbers work and where they are.

**The Riemann Hypothesis is still a mystery because even though many of these special points have been found on the line, no one has been able to prove that all of them are on the line.**

SOLVING THIS WOULD BE A HUGE BREAKTHROUGH IN MATH!

# Goldbach Conjecture

The Goldbach Conjecture is a very old and interesting math puzzle. It's named after a man named Christian Goldbach who thought of it a long time ago.

Here's the idea:

You know about even numbers, right? Even numbers are numbers like 2, 4, 6, 8, 10, and so on. They're all numbers that you can split into two equal parts without any leftovers.

The Goldbach Conjecture says that every even number that is bigger than 2 can be written as the sum of two prime numbers.

Now, prime numbers are special numbers that can only be divided by 1 and themselves without any leftovers. Some examples of prime numbers are 2, 3, 5, 7, and 11.

So, let's try this with an example:

Take the even number 8. According to the Goldbach Conjecture, we should be able to write 8 as the sum of two prime numbers. And we can! 8 = 3 + 5 (both 3 and 5 are prime numbers).

Another example: take the even number 10. We can write 10 as 5 + 5, and both 5s are prime numbers.

Goldbach's Conjecture says this should work for every even number bigger than 2, but no one has been able to prove this for all even numbers yet. It's like a big math mystery that mathematicians are still trying to solve!

# Twin prime conjecture

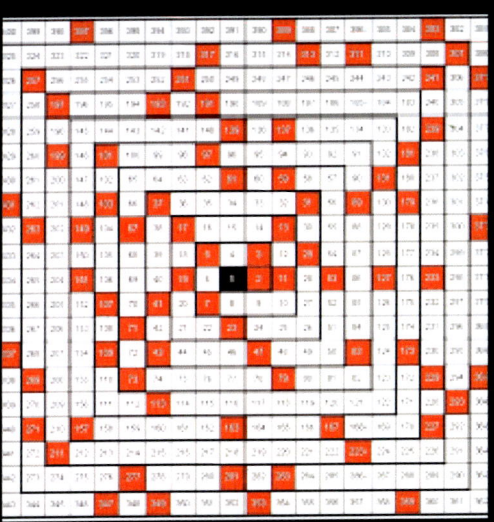

Let's talk about the Twin Prime Conjecture.

You know what prime numbers are, right? They are special numbers that can only be divided by 1 and themselves. Examples of prime numbers are 2, 3, 5, 7, 11, 13, and so on.

Now, let's talk about twin primes. Twin primes are pairs of prime numbers that are just 2 numbers apart.

For example:

3 and 5 are twin primes because they are both prime and the difference between them is 2.

11 and 13 are twin primes for the same reason.

17 and 19 are another pair of twin primes.

The Twin Prime Conjecture is a puzzle that mathematicians are trying to solve. It asks whether there are infinitely many pairs of twin primes. This means, as you count up higher and higher, will you keep finding pairs of prime numbers that are only 2 numbers apart, forever?

So far, mathematicians have found lots and lots of twin primes, but they haven't been able to prove that you can always find more twin primes no matter how high you go. It's like a mystery in math where we know some of the clues but haven't found the final answer yet.

# Collatz Conjecture

*Let's talk about the Collatz Conjecture, which is like a fun math game with numbers.*

## Here's how the game works:

1. Pick any positive number you like. Let's call this number your "starting number."

2. If your number is even (like 2, 4, 6), you divide it by 2.

3. If your number is odd (like 1, 3, 5), you multiply it by 3 and then add 1.

4. Take the new number you get and repeat the steps above.

The Collatz Conjecture says that no matter what number you start with, if you keep following these steps, you'll always eventually reach the number 1.

Let's try an example with the starting number 6:

6 is even, so we divide by 2 and get 3.

3 is odd, so we multiply by 3 and add 1, getting 10.

10 is even, so we divide by 2 and get 5.

5 is odd, so we multiply by 3 and add 1, getting 16.

16 is even, so we divide by 2 and get 8.

8 is even, so we divide by 2 and get 4.

4 is even, so we divide by 2 and get 2.

2 is even, so we divide by 2 and get 1.

And we reached 1!

Mathematicians have tested lots of numbers and it always seems to work, but no one has been able to prove it for every possible number yet. So it's a big, interesting mystery in math!

Fun fact: 27 takes 111 steps to complete its Collatz sequence

# Multiplying Large Numbers

Let's learn a cool trick to multiply big numbers that are close to 100, like 98 and 97. This trick will make you feel like a math magician!

Imagine you want to multiply 98 and 97. Here's how you can do it quickly and easily:

Subtract each number from 100:

Take 98 and subtract it from 100:

100−98=2.

Take 97 and subtract it from 100:

100−97=3.

Multiply the two differences we found:

2×3=6.

Subtract the sum of the differences from 100:

Add the differences together:

2+3=5.

Subtract this sum from 100:

100−5=95.

Combine the results:

Put together the 95 and the 6 to give us 9506.

So,

98×97=9506!

## Why This Works

This trick works because of some neat properties of numbers and how they interact. When you subtract the numbers from 100, you're essentially breaking down the multiplication into simpler steps.

This trick shows how fascinating and beautiful math can be! With just a few simple steps, you can multiply large numbers quickly and impress your friends and family with your math skills.

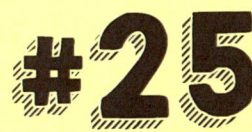

# Magic Squares

## What is a Magic Square?

A magic square is a special kind of puzzle that's like a magic trick with numbers. It's a grid (like a box) filled with numbers where each row, column, and diagonal adds up to the same number. That number is the magic number!

|   |   |   |
|---|---|---|
| 2 | 9 | 4 |
| 7 | 5 | 3 |
| 6 | 1 | 8 |

## How Does It Work?

### Grid with Numbers

Imagine a 3x3 grid (a square with 3 rows and 3 columns).

You fill this grid with numbers, like this:

```
4 9 2

3 5 7

8 1 6
```

### Magic Sum

Now, add up the numbers in each row, column, and diagonal:

- Row 1: 4 + 9 + 2 = 15

- Row 2: 3 + 5 + 7 = 15

- Row 3: 8 + 1 + 6 = 15

### Magic Number

In this example, the magic number is 15! That means every line you draw (across, up and down, and diagonally) will always add up to 15.

## Why is it Magic?

Magic squares are like puzzles because you can make different squares with different numbers, but they all have the same magic number. It's like a cool trick with numbers that always works out the same way.

## Fun Fact:

There are bigger magic squares, like 4x4 grids (4 rows and 4 columns) or even 5x5 grids! They work the same way—every row, column, and diagonal adds up to the same number.

# TRY IT YOURSELF

You can create your own magic square! Just remember, all the numbers have to add up to the same magic number in every direction. It's a fun way to play with numbers and see how they fit together like a puzzle!

Magic squares are a neat way to see how numbers can make patterns that are surprising and fun!

# Math Tricks

*Here are a few fun math tricks that you can easily learn and impress your friends with.*

## Trick 1: The 9 Times Table Trick

Have you ever noticed how the 9 times table has cool pattern? Here's a simple way to remember it:

1. Write down the numbers 0 to 9 in order:

   - 0, 1, 2, 3, 4, 5, 6, 7, 8, 9

2. Next to them, write the numbers 9 to 0 in reverse order:

   - 9, 8, 7, 6, 5, 4, 3, 2, 1, 0

Now, read the pairs of numbers together, and you get:

- 09 (which is 9 x 1)

- 18 (which is 9 x 2)

- 27 (which is 9 x 3)

- 36 (which is 9 x 4)

- 45 (which is 9 x 5)

- 54 (which is 9 x 6)

- 63 (which is 9 x 7)

- 72 (which is 9 x 8)

- 81 (which is 9 x 9)

- 90 (which is 9 x 10)

**Cool, right?**

## Trick 2: The Magic Number 1089

This is a fun trick that works every time!

1. Think of a three-digit number where the first and last digits are different (like 421).

2. Reverse the number (124).

3. Subtract the smaller number from the bigger number (421 - 124 = 297).

4. Now, reverse the result (792).

5. Add the result to the reversed number (297 + 792).

The answer will always be 1089!


## Trick 3: Multiplying by 11

Here's an easy way to multiply any two-digit number by 11.

1. Take a two-digit number (let's use 34).

2. Separate the digits (3 and 4).

3. Add the two digits together (3 + 4 = 7).

4. Place the sum in between the two digits of the original number (3 _ 4 becomes 374).

So, 34 x 11 = 374.

Note: If the sum of the digits is more than 9, just carry over the extra. For example:

- 57 x 11: 5 + 7 = 12, so put 2 in the middle and add 1 to 5: 627.

# Math Tricks continued

## Trick 4: The Disappearing Dollar Trick

This is a fun puzzle rather than a trick, but it's great for thinking!

Three friends go to a restaurant and share a meal costing $30. Each pays $10. The waiter realizes there's a mistake and the meal should have cost $25, so he gives $5 back. The friends decide to take $1 each and give $2 as a tip. Now, each friend has paid $9 (because they got $1 back from the $10, they originally paid), so $9 x 3 = $27. Plus, the $2 tip makes $29. Where did the missing dollar go?

The trick here is in how the math is being added up. There is no missing dollar. They paid $27, which includes the $25 for the meal and the $2 tip.

## Trick 5: Mind-Reading Number Trick

1. Think of a number (let's use 5).

2. Double it (5 x 2 = 10).

3. Add 8 (10 + 8 = 18).

4. Divide by 2 (18 / 2 = 9).

5. Subtract the original number (9 - 5 = 4).

No matter what number you start with, the answer will always be 4!

These tricks are a fun way to play with numbers and see some of the patterns and surprises in math!

# #28
# Perfect Numbers

A perfect number is a number that is equal to the sum of its proper divisors (excluding itself). For example, the number 6 is perfect because its divisors are 1, 2, and 3, and 1 + 2 + 3 = 6.

Imagine you have a bunch of friends, and you want to share something with them equally. Let's say you have 28 pieces of candy. You want to see if you can share these candies with your friends in such a way that everyone gets the same amount, and there's nothing left over.

A perfect number is a special kind of number that can be split up exactly like this, using its own pieces. Here's how it works:

First, find all the numbers that can divide the perfect number without leaving any pieces left over. These are called divisors.

Add up all these divisors (but don't include the number itself).

If the sum of these divisors equals the original number, then that number is perfect!

## Let's see an example of this:

The number 28

Find the divisors of 28 (excluding 28 itself):

1 (because 28 divided by 1 is 28)

2 (because 28 divided by 2 is 14)

4 (because 28 divided by 4 is 7)

7 (because 28 divided by 7 is 4)

14 (because 28 divided by 14 is 2)

Add these divisors together: 1 + 2 + 4 + 7 + 14 = 28.

Since the sum is the same as the original number, 28 is a perfect number

Since the sum is the same as the original number, 28 is a perfect number!

## So, perfect numbers are like magical numbers that can be split up perfectly into their own pieces!

Perfect!

# A fun math game.

Choose a number

Example: 12

Spell it out in English T-W-E-L-V-E

Count the letters – in this case it is 6

Spell it out in English S-I-X

Count out the letters

Keep going and see how long It takes you to get to 4 (it will always eventually get to 4)

Once you get to four it is impossible to get any other number except 4!

Do this with a friend and see who takes the longest to get to four.

#30 The final word count inside this book including this last fact is:

9603 words including numbers
1249 non-repeated words
50 943 characters with spaces
41 782 characters without spaces
624 paragraphs

**This includes footnotes and textboxes. It does not include speech bubbles or any pages after this one.**

ISBN: 978-1-7637973-1-4

First Printed in 2024
©Copyright 1000 Tales.

All rights reserved.
This book or any portion thereof may not be reproduced, stored in or introduced in a retrieval system, or transmitted, in any form or by any means without the express written permission of 1000 Tales except for the use of brief quotations in a book review.

Written by
Taarun Krushanth

Illustrated by
Ameera Karimshah and Atiya Karimshah

We would like to acknowledge the Traditional Custodians of the continent of Australia. Whose cultures are amoung the oldest living cultures in human history and whose languages and knowledge have infused and inhabited this land for millennia.

We recognise their continuing connection to the land, waters and culture and we pay our respects to their Elders past, present and emerging

# OTHER BOOKS BY TAARUN KRUSHANTH:

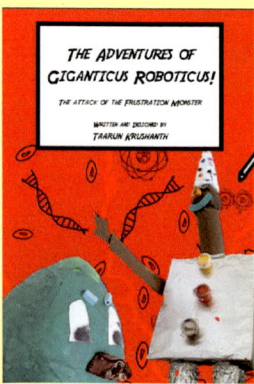
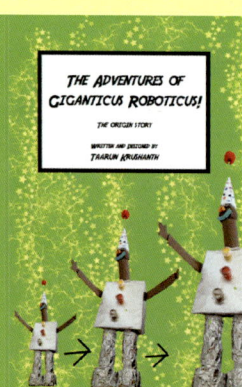

Coming SOON: **MATH KID 2!** and **THE ADVENTURES OF GIGANTICUS ROBOTICUS #5!**

www.ingramcontent.com/pod-product-compliance
Lightning Source LLC
LaVergne TN
LVRC080028250326
834741LV00018B/155